RAFTER BEEKEEPING

Sustainable Management with
Apis dorsata

I0168484

Eric Guerin

ISBN 978-1-913811-01-3

Jointly Published in 2021 by the International Bee Research Association, Monmouth (UK) and Northern Bee Books, Hebden Bridge (UK)

Available from the IBRA Bookshop: www.ibra.org.uk

and

Northern Bee Books: www.northernbeebooks.co.uk

Front cover: A rafter in Cambodia. Photo: Eric Guerin
Back cover: *Apis dorsata* worker fanning her nest. Photo: Eric Guerin

Cover design and artwork by DM Design & Print

IBRA Proof Editor - Stuart A. Roberts

Eric Guerin is a French biologist specialized in Asian native honey bee conservation and sustainable beekeeping. Based in Southeast Asia since 2008, he has made a special effort to document the work of honey hunters and rafter beekeeping communities, diffusing sustainable honey collection practices, introducing small-scale beekeeping as part of organic agriculture development, raising awareness on native honey bee conservation, and supporting optimization of forest honey value chains.

RAFTER BEEKEEPING

Sustainable Management with
Apis dorsata

Eric Guerin

IBRA
INTERNATIONAL BEE
RESEARCH ASSOCIATION

NBB

TABLE OF CONTENTS

1. INTRODUCTION

The giant honey bee, *Apis dorsata F.*, is a single-comb, open-nesting species of honey bee. Its distribution area covers much of southern Asia, including parts of Pakistan and all the way to the eastern part of the Indonesian chain of islands.

Because this species open nests, it has not evolved to live in dark cavities (Jack, Lucky, & Ellis, 2015). In addition, the colonies are defensive and demonstrate seasonal migration. Therefore, the domestication of this species, in the sense of the western honey bee *Apis mellifera L.* or the eastern honey bee, *Apis cerana F.*, is often thought to be impossible (Tan, 2004; Tan, Chinh, Thai, & Mulder, 1997; Chinh, Minh, Thai, & Tan, 1995).

Attempts to induce *Apis dorsata* to remain in a hive have failed (Figure 1) (Koeniger N, Koeniger G, Tingek, 2010; Crane, 1999; Thakar, 1973).

Figure 1: Apis dorsata colonies in experimental open fronted hives. Mysore state, India (Thakar 1973). Source: Crane, 1999.

However, in several locations of Southeast Asia, locals have used tree-poles, or rafters, positioned at a slight angle and low to the ground, that mimic tree branches, in order to attract migrating swarms of *Apis dorsata* (Petersen, 2012; https://www.beesunlimited.com). This traditional practice, first reported in 1902 by Fougères (Crane, Luyen, Mulder, & Ta, 1993), and commonly referred to as "Rafter Beekeeping," allow beekeepers to harvest honey two or three times from the same colony per season without harming the bees (Tan et. al., 1997).

As they are typically placed near the ground, rafters also allow much safer honey harvesting for honey collectors who are exposed to personal risk when attempting to reach colonies nesting far off the ground (Petersen, 2012; Waring & Jump, 2004) (Figure 2).

Figure 2: Apis dorsata honey hunter in Sumbawa, Indonesia. Photo Cooper Schouten.

This method of sustainable management of *Apis dorsata* can be introduced into suitable environments in parts of the tropics and subtropics (Petersen & Reddy, 2016).

Intended to serve as a training manual the following has been developed on the basis of related literature as well as personal observations by the author. It aims to provide the essential requirements for the implementation of rafter beekeeping and presents guidelines for rafter construction and management as a means of sustainable harvesting. Nevertheless, it should not be considered as a substitute for practical training by an experienced rafter beekeeper, at least in the first stages of implementation. As is the case for any beekeeping activity, experience is a key factor of success.

2. WHAT IS RAFTER BEEKEEPING?

Rafter beekeeping refers to the manufacture of purpose-made nesting sites to attract migrating swarms of the giant Asian honey bee (*Apis dorsata*) (Petersen, 2012).

A rafter is the trunk or branch of a tree, around 2 m in length and 10 to 20 cm in diameter supported by two vertical poles (Figure 3). Resembling a branch of a tree, a rafter attracts *Apis dorsata* to build a nest beneath it (Chinh et al., 1995).

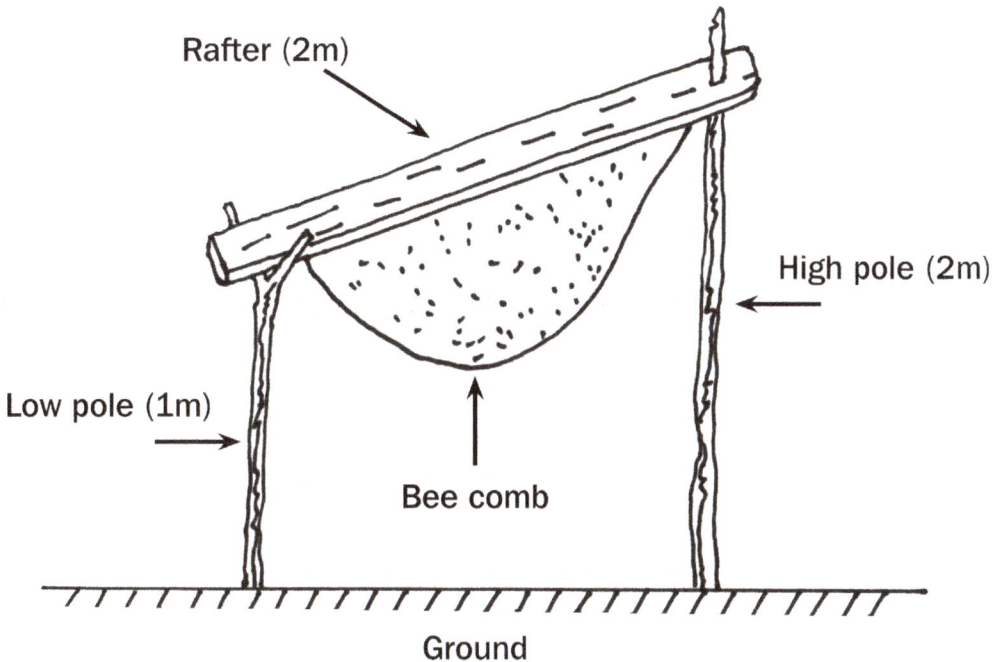

Figure 3: Diagram of a rafter and Apis dorsata comb. Source: Chinh et al., 1995.

Because the tree poles are placed at an angle equal to 30 degrees, resembling the rafters, or structural roof components of a building, the method is called rafter beekeeping (Petersen, 2012; Chinh et al., 1995). And because of the rafter's slope the bees store the majority of the honey at the upper end of the rafter resulting in a "honey head" that may be sustainably harvested with no detriment to the brood (Petersen, 2012; Chinh, et al., 1995).

Rafter beekeeping is, and has been for many hundreds of years, practiced in several locations across Southeast Asia (Table 1) (Petersen, 2012; Petersen, 2010; Mulder, Heri, and Wickham, 2000; Tan et al., 1997).

Table 1: Location and local names of rafter beekeeping in South East Asia.
Modified from Petersen, 2012.

Location	Local name	Season	Authors
Indonesia, Kapuas Lakes, West Kalimantan	Tikung	Dec-March Rainy season	Petersen, 2010; Crane et al., 1993
Cambodia, Koh Kong	Bong kong	June-Sept Rainy season	
Cambodia, Siem Reap	Bong kong	Oct-April Dry season	Petersen, 2005; Jump & Waring, 2004; Waring & Jump, 2004
Indonesia, Bangka Island & southern Sumatra	Sunggau		Nurtjahya, 2012; Purwanto, Hadisoesilo, Kasno, Koeniger, Lunderstedt, 2001
Indonesia, Sulawesi	Tingku,		Hadisoesilo, 2001
India	Attraction planks		Mahendre, 1997
Vietnam, U Minh forest, Mekong delta	Gac-keo	2 seasonal harvests	Crane et al., 1993; Tan et al., 1997; Chinh et al., 1995
Thailand, Songkhla lake basin	Bang Kad	June-December (dry and rainy seasons)	Bendem-Ahlee, Kittitornkool, Thungwa & Parinyasutinun, 2014, 2015; Chuttong, Somana, & Burgett , 2019

Petersen (2012) distinguished two groups of rafter designs "based on the season in which the bees are producing honey in the area; dry season, as in Siem Reap, Cambodia, and wet season as in Kapuas/Danau, Sentarum, West Kalimantan, Indonesia."

Petersen characterized a dry season rafter as less elaborate than a rainy season one. "Most frequently dry season rafters are simple round logs, without any shaggy bark, supported at either end by a forked pole driven into the ground or a natural fork in a small tree. In contrast, the rainy season rafters, are hewn to shape, being concave on the upper surface, much like a rain gutter, to drain off rain water so as not to seep into the comb and convex on the lower surface (simulating a smooth branch)." (Petersen, 2012).

Petersen's description, however, does not actually appear to be so systematic. Rafter beekeepers in the Mekong delta of Vietnam, and in West Kalimantan in Indonesia, have developed more sophisticated rainy season rafters, while rafter beekeepers in southern Thailand (Songkhla) and southern Cambodia (Koh Kong) have adopted the simple dry season design (Figure 4). Basic rafters also work in a rainy season context.

Figure 4: Simple "dry season design" rafters in southern Thailand (upper) and more sophisticated "rainy season type" rafters in southern Vietnam (lower) (see paragraph 5:1 "Gac-Keo"). Photos: Eric Guerin.

3. BASIC REQUIREMENTS FOR RAFTERING

Stephen Petersen identified five essential criteria for a successful implementation of rafter beekeeping, some structural and others more concerned with environmental and cultural factors (Petersen, 2012). The five criteria include:

1. Bees present in the local environment.
2. A lack of "natural" nesting sites, such as tall trees, water towers and cliffs, which will force the bees to nest on the purpose-made rafters.
3. An abundance of bee-friendly floral resources available at least on a seasonal basis.
4. A widely recognized tenure or ownership of the rafters by individuals enforced by community laws and culture.
5. Locally available, long-lasting, sustainable materials from which to construct the rafters.

Recent experiences have revealed the importance of an additional criterion; a minimum level of clarity on the tenure and management of the land where the rafters will be set up.

3.1 Presence of bees in the environment

The purpose of the rafter is to attract bee swarms, thus, having a rafter occupied in an area where bees are not already present is quite unlikely.

3.2 Lack of "natural" nesting sites

The lack of natural nesting sites might be due to the small-sized branches of the secondary growth forest, as seen around Siem Reap, Cambodia, the vertical nature of the branches or the unsuitable shaggy or papery and loose bark texture of the main tree species, *Melaleuca leucadendron* in southern Cambodia, Thailand and Vietnam (Figure 5) (Nagir et al., 2016; Petersen & Reddy, 2016; Saberioon, Mardan, Nordin, Alias, & Gholizadeh, 2010; Petersen, 2005; Reddy & Reddy, 1989).

Although wild nests can be found at low height and even near ground level (Nagir, Atmowidi, & Kahono, 2016), *Apis dorsata* usually establishes its single-comb colony high up in tall trees (Sihag, 2017; https://www.beesunlimited.com). As a consequence, rafter beekeeping is particularly adapted to environments where the only suitable nesting sites for these bees are the rafters (Jack et al., 2015; www.beesunlimited.com).

Nevertheless, trials in several provinces of Cambodia by Bees Unlimited and Eric Guerin showed that rafters surrounded by potential nesting trees, could also be colonized by *Apis dorsata* colonies. Rafters close to forest areas were sometimes colonized only one week after construction.

Figure 5: Typical environment with a lack of "natural" nesting sites. Upper, melaleuca forest in southern Thailand; lower secondary forest regrowth near Siem Reap, Cambodia. Photos: Eric Guerin.

3.3 Abundance of bee friendly floral resources

In any beekeeping project, the abundance of bee friendly floral resources, in terms of both nectar and pollen, is an essential requirement. Substantial areas of bee friendly forage must be present at least on a seasonal basis (Petersen, 2012).

As stated by Petersen & Reddy (2016) "it is important to establish the honey potential of an area by observations of the phenology (bloom time) of the bee friendly plants. Locals typically have a reasonable understanding of which flowers are most attractive to bees and their respective blooming season; they are also aware of the seasonal migrations in response to blooms that are typical of *Apis dorsata*". In addition, although *Apis dorsata* are able to fly several kilometers on foraging trips, they are much more efficient at close range (Petersen, & Reddy, 2016).

Proximity to pollen sources should not be neglected as *Apis dorsata* prefers to build its nest in close proximity to the pollen source (Ibrahim, Siva, Balasundram, Abdullah, Sood, Alias, Mardan, and Saberioon, 2012). The presence of a water source, throughout the flower bloom period, should also be insured as colonies might abscond prior to the end of the nectar flow in case of water scarcity (personal observation from the author in Mondulkiri, Cambodia).

3.4 A widely recogonized tenure or ownership of the rafters by individuals enforced by community laws and culture

As emphasized by Petersen and Reddy (2016), wild bees are seen as a common, free resource to be exploited by whoever discovers them. For rafter beekeeping to be successful and sustainable, a "sense of ownership" must be present that is locally enforced by surrounding communities (Figure 6). Otherwise, outsiders, with no respect for the ownership, will frequently harvest the honey, often in an unsustainable manner such as taking the brood or killing the bees to obtain the honey.

Theft prevention should be taken into account in the implementation of rafter beekeeping. Even though it is obvious that the rafters are man-made, the bees themselves are often seen as a "wild resource" to be harvested on a first-come basis (Petersen, & Reddy, 2016). "It does require some special skills to harvest honey without massive stinging attacks but determined thieves can circumvent this problem by using insecticides." (Petersen, & Reddy, 2016).

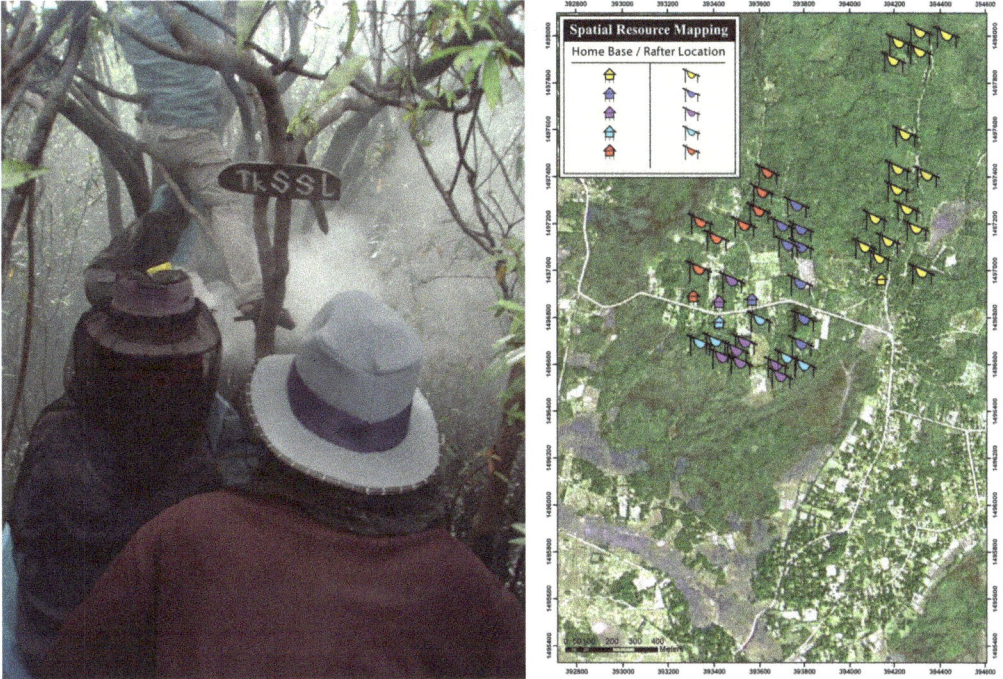

Figure 6: Ownership of a rafter indicated by an owner's mark and mapping location in West Kalimantan, Indonesia. Source: Stephen Petersen, 2012.

3.5   Clarity on the tenure and management of the land

In order to avoid nasty surprises such as the destruction of occupied rafters by land clearing or bush fires as a result of slash and burn agriculture (Figure 7), the ownership, as well as the management plans of the land for upcoming seasons, should be assessed prior to building rafters. This point is particularly important in the case of an introduction of rafter beekeeping to other communities. The frustration generated by the destruction of hours of work might discourage villagers to further explore the technique.

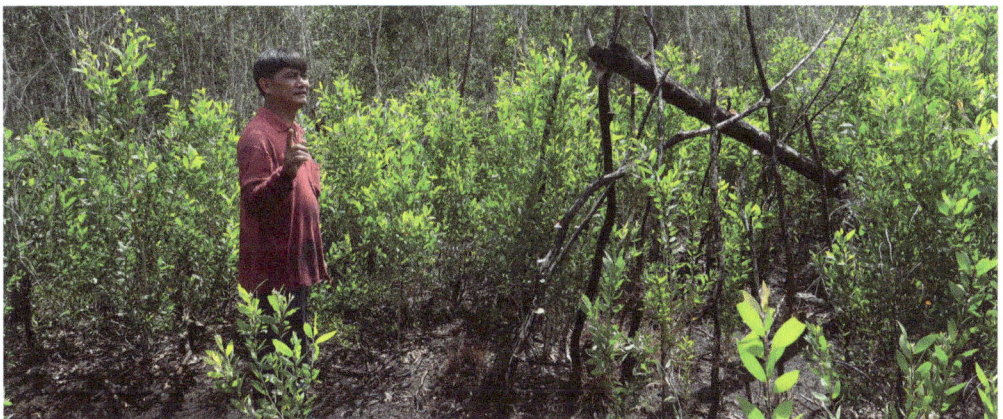

Figure 7: A rafter destroyed by forest fire (southern Thailand). Photo: Eric Guerin.

4. RAFTER DESIGN AND CONSTRUCTION

Though a rafter is quite a simple construction, basically a trunk or a branch of a tree about 2 meters in length supported by two vertical poles with a slope of around 30 degrees, its chances of being effectively occupied by a bee swarm depends upon compliance with several parameters described in the next section.

4.1 Wood selection

Main rafter beam

Based on rafter beekeepers experience and literature related to *Apis dorsata* natural nesting sites, it seems that a wide variety of wood types can be used for rafter construction. Nagir et al., (2016) found 34 species belonging to 27 genera in 17 families of plants as nesting sites of giant honey bees. And in southern Thailand, Chuttong et al. (2019), identified 6 tree species used to build rafters.

Nevertheless, the following common-sense guidelines help in making appropriate selection of wood types:

▸ Sturdy and woody branches, smooth-barked and hard to peel branches, long lasting wood, should be preferred (Figure 8) (Nagir et al., 2016).

▸ On the contrary, wood with papery and loose, shaggy or unevenly grooved barks (Sihag, 2017) as well as wood repellent to bees (local knowledge) should be avoided.

▸ Wood should be locally available and shouldn't pose a threat to the environment by its utilization (Petersen, 2012).

Figure 8: Though Apis dorsata colonies prefer smooth surfaces to hang their nests on, they seem to have a certain degree of tolerance for grooved barks (occupied rafter in southern Thailand). Photo: Eric Guerin.

Bamboo can be used since its lightness facilitates transportation, but bamboo rafters have a short lifespan and must be replaced often (P.H. Chinh, personal communication, November 13, 2019).

Though *Melaleuca cajuputi Powell* is considered as inadequate by Thai and Cambodian beekeepers due to its papery and loose shaggy bark (Chuttong et al., 2019), it is the most preferred wood by Vietnamese beekeepers, after peeling the bark, due to its longer lifespan (P.H. Chinh, personal communication, November 13, 2019).

Vertical poles

Wood selection for the vertical poles is not as crucial as it is for the main rafter beam. As long as the vertical poles can bear the weight of the rafter plus the bee colony, most types of locally available wood can be used.

One or both vertical poles can be avoided by making best use of the vegetation (Figure 9). The rafter can be directly mounted between 2 small trees.

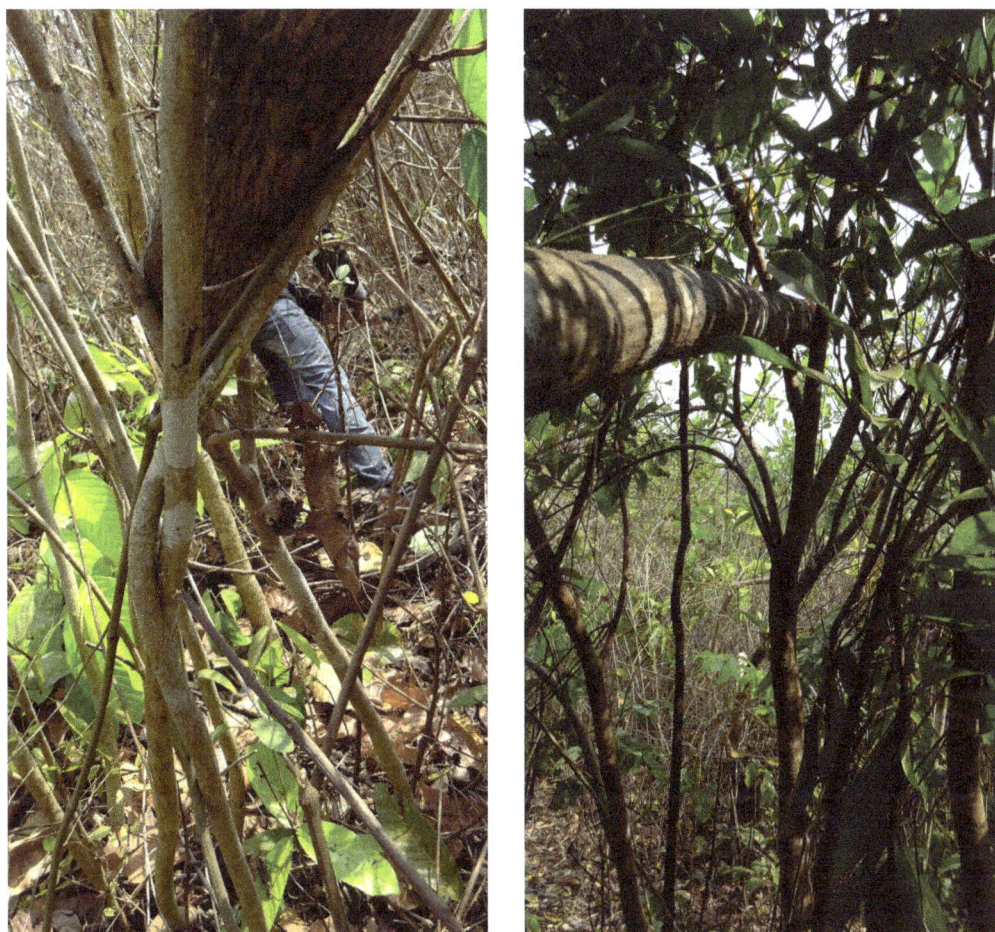

Figure 9: The main rafter beam can be directly mounted between 2 small trees (Mondulkiri, Cambodia). Photos: Eric Guerin.

4.2 Rafter length and circumference

Rafter length

The main rafter beam should be long enough to allow a full development of the colony (Tan, 2004). Experiments in Vietnam showed that the lengths of rafters above the maximum measured comb length, approximately 1.60 meters, had no influence on bee occupation (Tan, 2004). Lengths in the range of 2.5–3.0 meters ensure adequate distance between the two vertical poles. All rafters measured by Chuttong et al. (2019) in southern Thailand averaged 2.58 (+/-) 0.2 meters.

Rafter diameter

The main rafter beam should at least be able to bear the weight of the colony (wax + brood + honey + adult bees) at its full development; 38kg (Tan, 2004).

The circumference of the rafter also seems to be a crucial parameter in the occupation of a rafter. Generally speaking, as bees tend to choose strong nest sites, the bigger the rafter, the higher the occupation rate (Tan, 2004; M.S. Reddy & Reddy, 1989). In their rafter assessment in Vietnam, Tan et al. (1997), found that the rafter occupation rate increased significantly with rafter's circumference.

However, as showed by Sihag (2017) the relationship between strength of support and occupation rate is positive up to a certain limit (Tan, 2004); 20 centimeters being the preferred diameter. Of the 440 nests assessed by Sihag, 48% (210) nested on branches with a diameter of 20 centimeter. 26% (115) on 25 centimeter diameter branches, 19% (85) on 15 centimeter diameter branches and 7% (30) on 10 centimeter diameter branches.

4.3 Rafter slope

The slope of the rafter should be compatible with the nesting requirements of *Apis dorsata* and inducing the bees to store most of their honey at the upper end of the rafter resulting in a "honey head" (Figure 10) that may be sustainably harvested with no detriment to the brood.

Figure 10: Because of the rafter's slope the bees store the majority of the honey at the upper end of the rafter resulting in a "honey head". Photo: Eric Guerin.

Apis dorsata can nest on supports with a slope up to 60 degrees (Nagir et al., 2016). Nevertheless, most natural nests are found on supports with an inclination from 0 to 45 degrees (Sihag, 2017).

An excessive inclination might not be attractive to the bees, while a colony nesting on a slightly too inclined rafter will store its honey (pink) above the pollen (yellow) and brood (blue) making it difficult to harvest honey while leaving the comb intact (Figure 11) (Petersen, 2012).

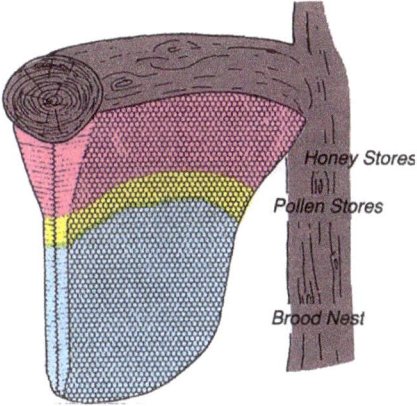

Figure 11: Apis dorsata nest on a horizontal branch: honey (pink) above the pollen (yellow band) and brood (blue). Source: Punchihewa 1994.

Rafter slope also affects the occupation of the nest sites by bees (Tan, 2004). Research conducted in Vietnam on 507 rafters (occupied and not occupied with bees) showed that rafters with an angle in the range of 27 to 40 degrees had the higher occupation rate (Tan et al., 1997).

4.4 Rafter height

Rafter poles are designed to obtain a rafter with a slope of around 30 degrees and conveniently placed at eye-level in order to provide easy access to the comb and in particular to the honey head (Figure 12) (https://www.beesunlimited.com).

In flooded areas, such as melaleuca forests, the height of the rafter should ensure that the bee comb is above rising water levels.

The upper end of the rafter can be carved with a notch to secure the rafter in place (Figure 13).

In Cambodia, the front pole is adjusted in order to stand above the vegetation (see 4.9 "Rafter location") (Figure 14).

Figure 12: The rafter is conveniently placed at eye-level in order to provide an easy access to the comb (Siem Reap, Cambodia). Photo: Eric Guerin.

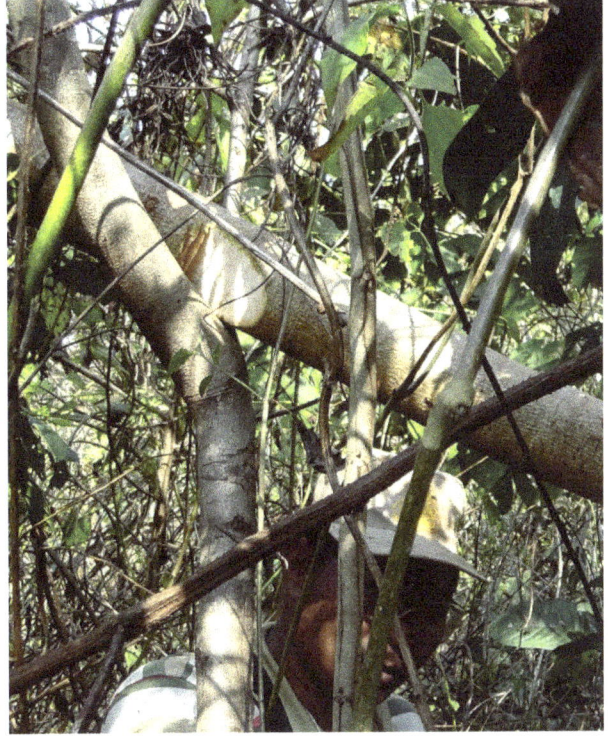

Figure 13: Notch at the upper end of the rafter to facilitate securing the rafter in place (Cambodia). Photos: Eric Guerin.

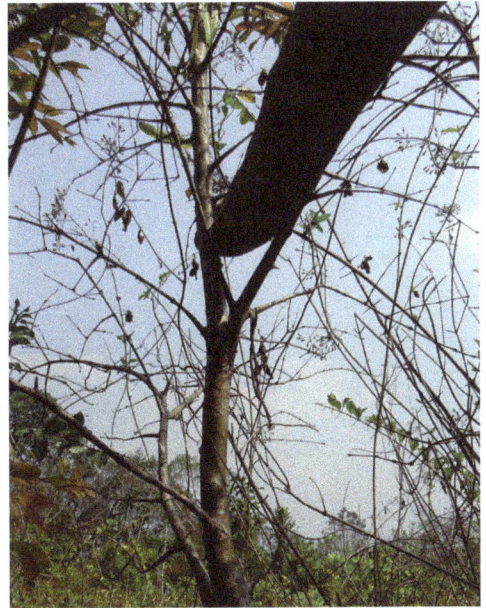

Figure 14: Rafter level adjustment according to vegetation height (Mondulkiri, Cambodia). Photo: Eric Guerin.

4.5 Shading

In areas where vegetation may be patchy, the rafter should be shaded by cutting or bending small branches over the rafter in a tent-like fashion (Figures 15 & 16) (Chuttong et al., 2019; Petersen, 2012; Chinh et al., 1995).

Figure 15: Bending small branches over a rafter (Mondulkiri, Cambodia). Photo: Eric Guerin.

Figure 16: Rafter shaded in Cambodia (upper) and Thailand (lower). Photos: Eric Guerin.

4.6 Bee flyaway and access

Branches in front of the rafter's upper end should be trimmed to allow the bees easy access to the nest (Figure 17) (Petersen, 2012).

Figure 17: Flyway or tunnel through the thick vegetation to allow the bees an easy access to the nest (Siem Reap, Cambodia). Photos: Alain Montane.

4.7 Underneath vegetation clearance

Grasses and small trees beneath the rafter which might touch the future comb should be cleared to prevent access to the nest by predators and to avoid damage to the comb (Figure 18).

Figure 18: Underneath vegetation clearance (Siem Reap, Cambodia). Photos: Eric Guerin.

To increase the chance of occupation by a swarm, rafters should be checked periodically and vegetation underneath, cleared. In Thailand, beekeepers remove nearby green ant or weaver ant nests prior to setting up a rafter.

4.8 Wax coating

The lower surface of the rafter should be coated with melted beeswax or rubbed with a block of bee wax (Figure 19) to simulate a previous occupation and enhance attractiveness (Figure 20). Investigations on arrival and settlement of *Apis dorsata* swarms in Nepal and India showed the interest of *Apis dorsata* scout bees for traces of combs from previous seasons (Woyke, 2017). In Vietnam, Tan (2004) showed that rafters with remnants of beeswax were four times more likely to be occupied by bee colonies than those without relics of old nests.

Wax coating is particularly important in areas where natural nesting sites such as tall trees might otherwise be favored by bees.

Figure 19: Rubbing a rafter with a block of beeswax (Mondulkiri, Cambodia). Photo: Eric Guerin.

Figure 20: Rafter with traces of combs from previous season. Photo: Stephen Petersen.

4.9   Rafter location

The choice of a good location is an essential parameter of successful rafter occupation.

As wild nests of *Apis dorsata* are usually protected by liana foliage (Nagir et al., 2016), rafter beekeepers typically choose a quiet open space surrounded by fairly dense vegetation or clear an open space by cutting down small trees. The surrounding vegetation protects the colony against external disturbance, which might result in colony absconding, and shades the comb, contributing to the thermoregulation of the brood. In northern Cambodia, rafter beekeepers consider the height of an elephant as the reference for optimal surrounding vegetation.

Rafter beekeepers emphasize the importance of an open space in front of the higher end of the rafter (Chuttong et al., 2019; Petersen, 2012; Tan, 2004; Chinh et al., 1995; Crane, 1990; Ruttner, 1988). In northern Cambodia, where rafters are set up in dense vegetation, beekeepers cut or bend the small trees in front of the upper end of the rafter over an area of a dozen meters. In Thailand, rafters are placed along the edge of the Melaleuca forest (Chuttong et al., 2019). In Vietnam, Tan et al. (1997) found that the "occupation was the highest (85% and 92% in dry and rainy seasons, respectively) for rafters with a diameter of greater than 25 meters of open space in front of them" (Figure 21).

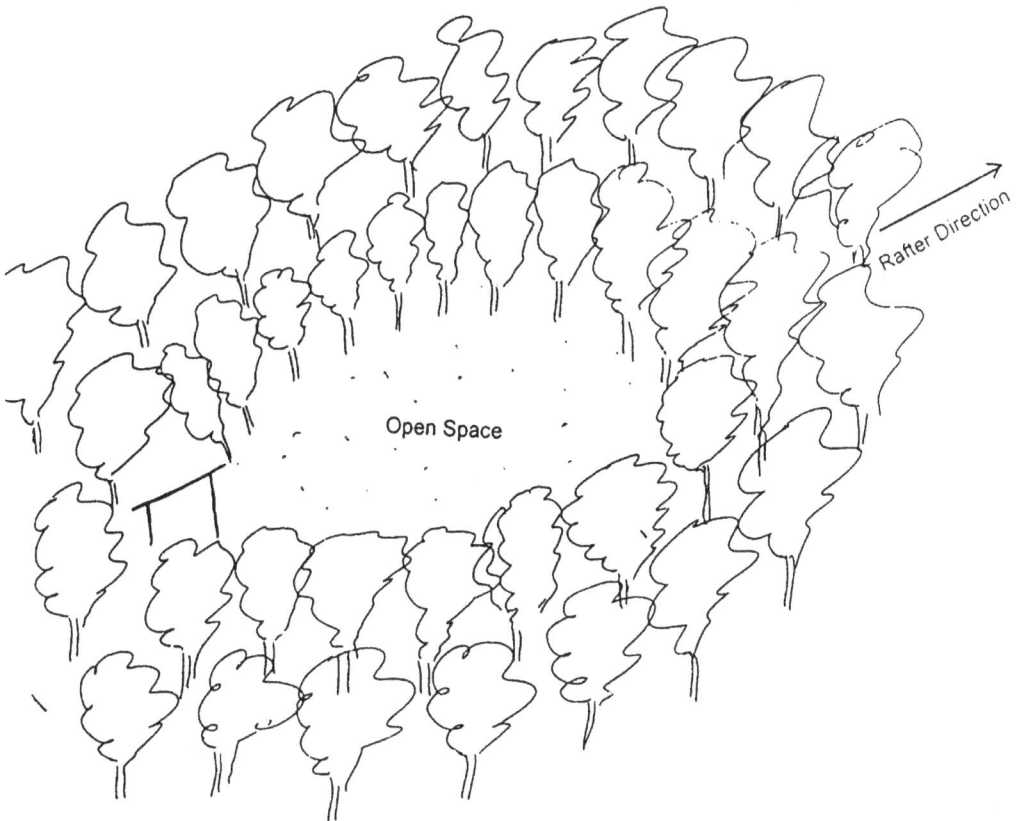

Figure 21: Open space in front of rafters (Vietnam). Source: Nguyen Quang Tan, 2004.

This open space might facilitate the bees' access to the nest, help them to avoid predators or facilitate the initial discovery of the site by scout bees (Tan, 2004). The explanation remains unclear, however, this criterion is crucial in "the occupation of the nest site by a bee swarm" (Tan, 2004)

In Thailand some rafters are even set up in grass lands (Figure 22).

Figure 22: Rafter in grass land (southern Thailand). Photo: Eric Guerin.

4.10 Rafter orientation

The compass orientation of the rafters has no effect on their occupation by a bee swarm (Tan, 2004; Tan et al., 1997).

4.11 Distance between rafters

At least for *Apis dorsata*, distance between rafters isn't an issue in terms of attractiveness as the subspecies is an aggregate nester (Figure 23).

The predominant concern, while considering the set-up of nearby rafters, is the avoidance of side attacks by neighboring nests during honey harvesting or colony monitoring (Mulder et al., 2000).

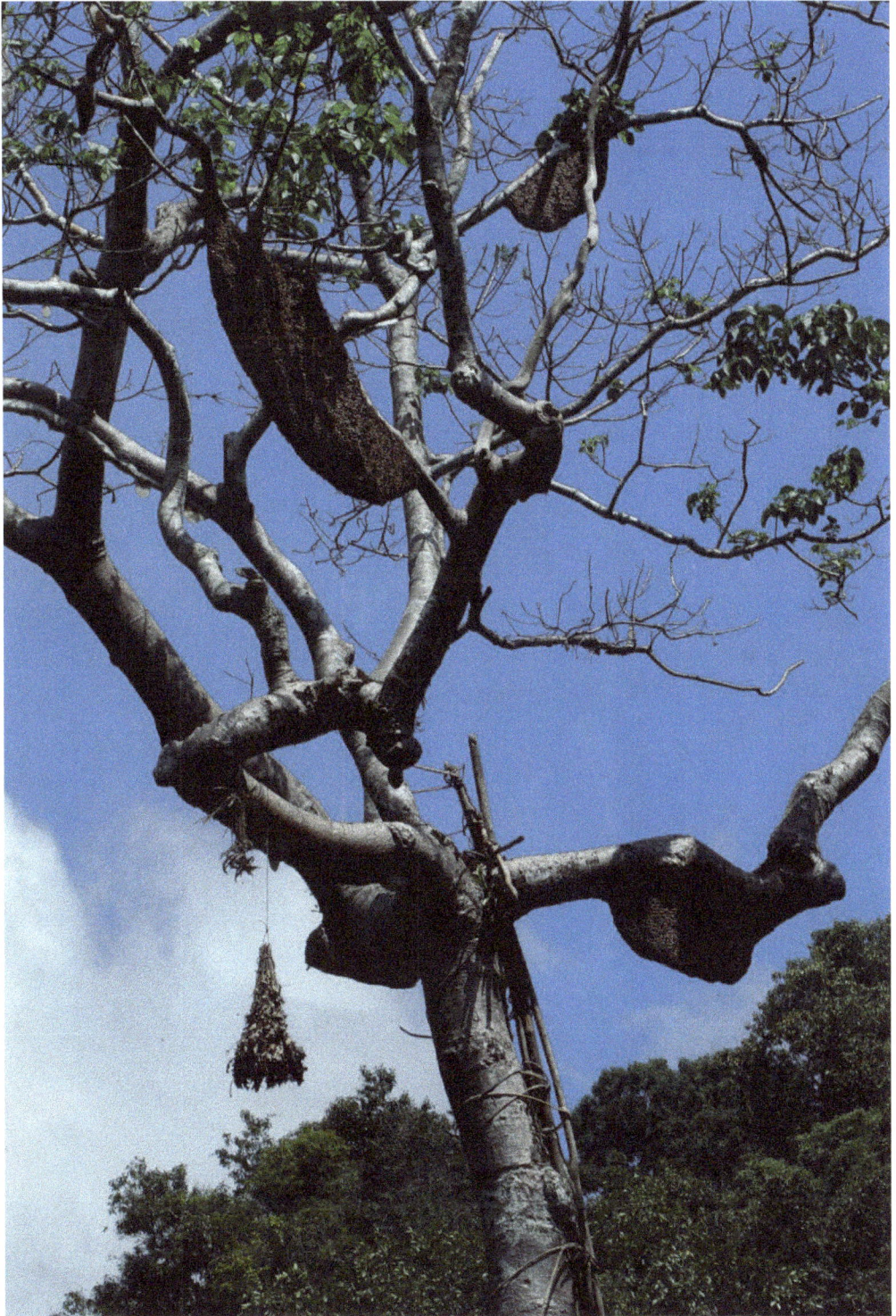

Figure 23: Apis dorsata colonies aggregation in a bee tree. Photo: Cooper Schouten.

4.12   When to install rafters?

To optimize the chances of occupation, rafters should be set up ahead of the migration of the bees to the area (Figure 24).

Figure 24: A rafter, one week after a bee colony settled (upper) and the same rafter 2 weeks later (lower) (Siem Reap, Cambodia). Photos: Alain Montane (upper) and Eric Guerin (lower).

5. ALTERNATIVE RAFTER DESIGNS

5:1 GAC-KEO (VIETNAM)

In Vietnam, beekeepers split tree trunks lengthwise, usually from *Melaleuca cajuputi*, in lengths between 1.8 and 2.2 meters with diameters of 10-20 centimeters to make two rafters (Figure 25). The outer part of each rafter is then peeled to remove the bark. Some beekeepers coat the curved side with beeswax (Tan, 2004; Chinh et al., 1995).

Figure 25: Newly built rafter (left) and rafter occupied by a bee colony (right) in Vietnam. Photos: Eric Guerin.

"A channel is often made along the flat side of the rafter to drain off rain water so that it does not seep into the comb. A rectangular or triangular hole is made at one end of the rafter. The rafter is supported by the higher pole with the hole slotting over the end of the pole (Figure 26). The lower pole supports the rafter with its V-shaped top. The curved side of the rafter faces downwards" (Chinh et al., 1995).

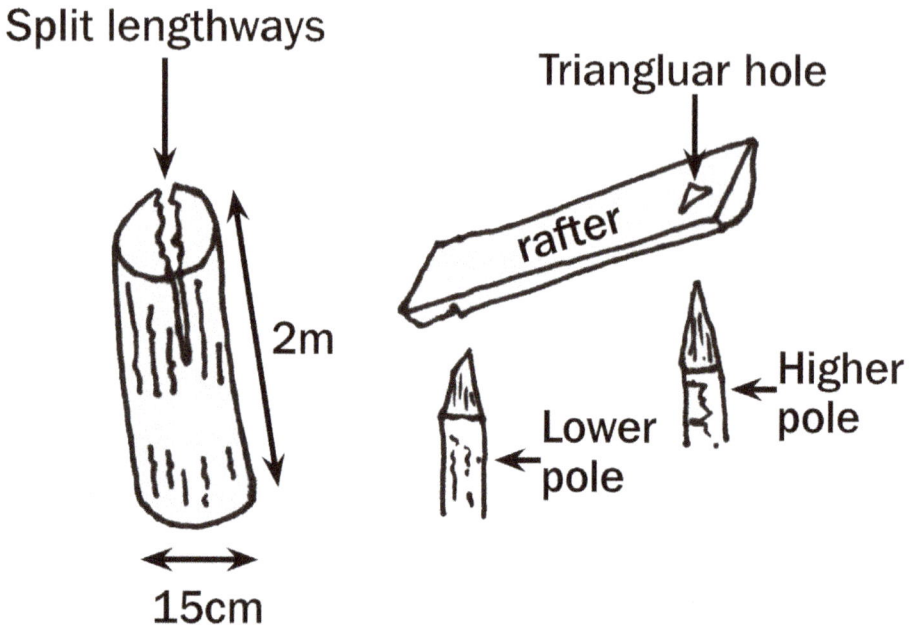

Figure 26: Rafters design in Vietnam. Source: Chinh et al., 1995.

5:2 "Tikung" (Indonesia)

The *Tikung* is an Indonesian rafter designed for rainy season rafter beekeeping. This technique documented by Petersen (2012) and Mulder et al. (2000) is used in the Kapuas Lakes region of West Kalimantan in Indonesia.

The *Tikung* design is made from a carved hardwood plank, approximately 0.8-2.5 meters long by 25-40 centimeters wide, commonly made of *Fagraea fragrans* or sometimes *Litsea sp* (Mulder et al., 2000). The upper surface has a concave shape, as in a rain gutter, allowing water to run off. The lower surface is rounded and smoothed to mimic a large branch (Figure 27) (Petersen, 2012; Mulder et al., 2000).

Figure 27: Tikung (rainy season rafters). Photos: Stephen Petersen.

Tikung are attached to tree branches in stunted, submerged forests (Mulder et al., 2000). Notched ends with pegs secure the *Tikung* in place (Figures 28 & 29) (Petersen, 2012; Mulder et al., 2000). A pole is sometimes horizontally attached about 2 meters below the *Tikung* to allow the beekeeper to stand on it while attaching and during harvesting (Mulder et al., 2000).

Figure 28: Setting up a Tikung (West Kalimantan, Indonesia). Photo: Stephen Petersen.

Though it takes a man more than one day to make a *Tikung* plank (Mulder et al., 2000), it has a much longer lifespan than Cambodian, Vietnamese or Thai rafters (Figure 30). According to Mulder et al. (2000) "*Tikung* planks can last over 2 generations (40 years) and can still be used after enduring a serious forest fire".

Figure 29: Tikung Rafters. Photos: Phy Bunthorn/Pact Cambodia Research Team, CFP (upper) and Stephen Petersen (lower).

Figure 30: Tikung rafters occupied by a bee colony (West Kalimantan, Borneo). Photo: Stephen Petersen.

5:3 Attraction planks (India)

In southern India, a method similar to rafters was developed by Muthappa (1979). Wooden planks, 1.5 meters long, 15 centimeters wide and 2.5 centimeters thick, and coated with bee wax were fixed on tall trees or rock faces where *Apis dorsata* colonies usually nested (Crane, 1999). At harvesting time, the planks with the bee nest were pulled down to the ground using ropes and a series of pulleys (Figure 31) (Crane, 1999). The lower part of the comb containing the brood and pollen was tied back to the plank and hoisted up again (Crane, 1999). In 1997, Mahendre improved the system to ease the honey harvest; after a plank was occupied and comb building initiated, one end of the plank was lowered slightly to form an angle where the bees would store honey on the upper most apex (Mahendre, 1997 in Petersen, 2012).

Figure 31: Attraction plank in southern India. Photo: D.B. Mahindre

6. HARVESTING AND PROCESSING HONEY AND WAX FROM RAFTERS

Compared to honey hunting on feral nests built on tall trees, honey harvesting from a rafter placed at ground level is much easier and safer for the beekeeper (Waring & Jump, 2004).

A visible honey head with fewer bees covering it, as well as an increased water collecting activity of the worker bees, are good indicators that the honey is ripe for harvest (Chinh et al., 1995).

Honey should be harvested during the day as it is far safer or more "bee friendly" than night harvest which results in the loss of many bees and possibly many more stings (Petersen, 2012).

In northern Cambodia, beekeepers prefer accessing the rafter from behind, which is the lower part, or from one side to avoid bee stings. This rule doesn't seem to be followed by Thai and Vietnamese beekeepers where beekeepers access rafters from the front or upper end (Figure 32).

Figure 32: In Thailand, beekeepers access the rafter by the front side. Photos: Eric Guerin.

The bees are driven away from the comb with the help of a "smoker" (Figure 33). Only cool-white smoke should be used. "Smoke, properly used, is a very effective deterrent to the bees and should protect the beekeepers from stinging incidents, provided they stay in the smoke" (Petersen, 2012). As much as possible, smoking the honey head should be avoided to prevent giving a bad smell and taste to the honey.

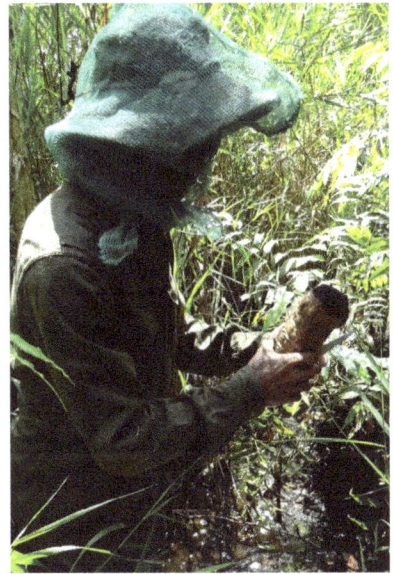

Figure 33: Traditional smokers in northern Cambodia (left) and southern Vietnam (right). Photo: Eric Guerin.

In Vietnam, beekeepers wear head protection, or bee veils, to prevent bee stings. Cambodian and Thai beekeepers do not take such precaution and approach rafters without any protective equipment. However, all beekeepers emphasize that the honey harvest should be completed quickly, in 5 to 10 minutes, in order to leave the rafter before the return of bees (Figure 34).

Figure 34: Checking a rafter in northern Cambodia. Photo: Alain Montane.

Only clean harvesting implements, such as knives and buckets, should be used when harvesting honey. The "bee bread," or pollen, stored just underneath the honey head, should be cut off prior to the honey head in order to limit the quantity of pollen mixed with the honey. As much bee bread as possible should be left to the bees. Only the honey head is cut off and around 20 percent of the honey should be left to the bees to allow for multiple harvests (Figure 35) (Chuttong, 2019; Petersen, 2012).

Some Vietnamese rafter beekeepers consider that cutting a part of the brood increases the honey yield of the next harvest. Though cutting a part of the brood would logically weaken the colony and consequently reduce its capacity to store honey, higher honey yields might be linked to the removal of queen cells which prevents swarming (Chinh et al., 1995). This practice, however, should be banned because of its potential negative effect on swarming, a natural reproduction process of honey bees. Swarming is crucial to the conservation of bee populations.

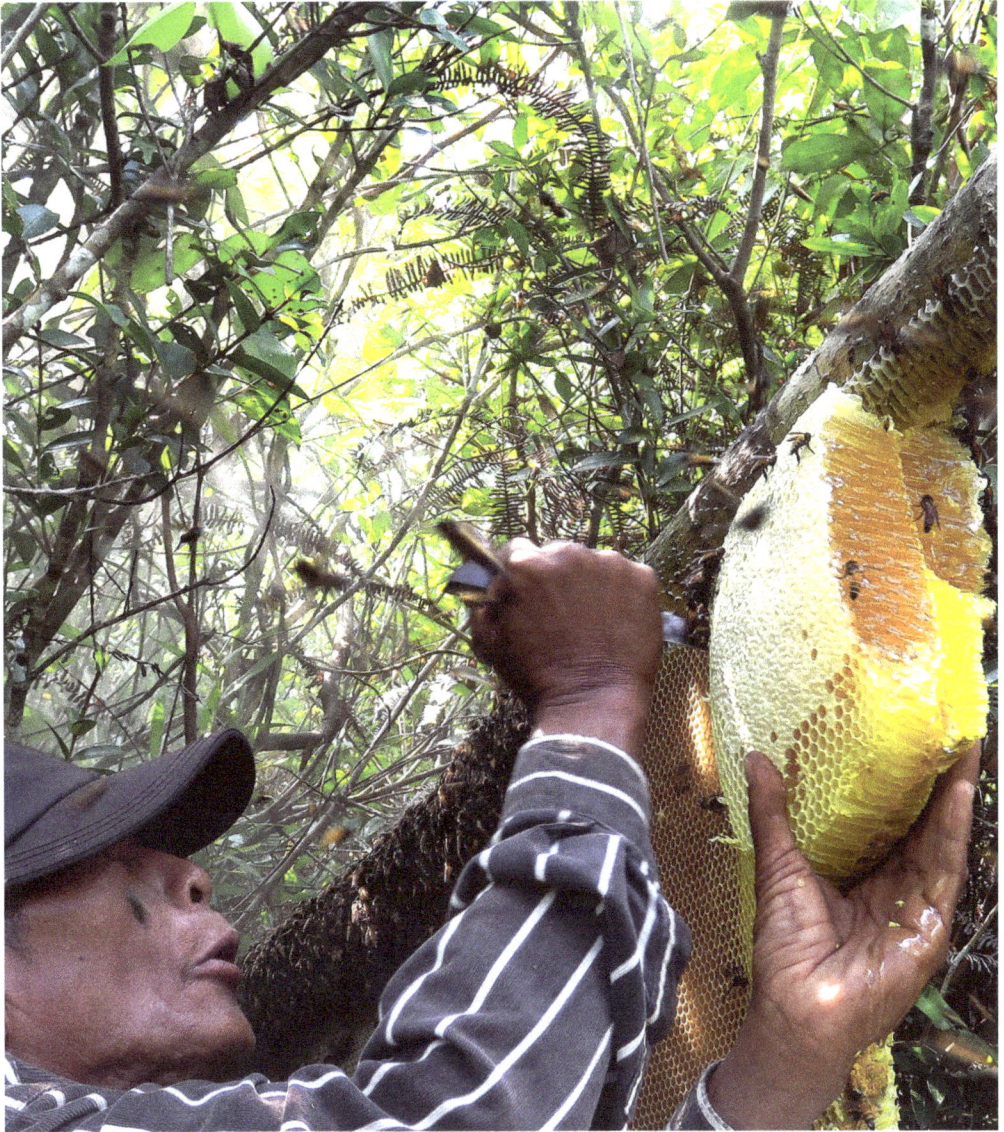

Figure 35: Sustainable harvesting form a rafter (Siem Reap, Cambodia). Photo: Eric Guerin.

High moisture, more than 22 percent, is common in *Apis dorsata* honeys. To limit the risk of fermentation, as much as possible, only the "ripe" or capped over, portion of the honey should be harvested and care should be exercised to avoid contamination by pollen or brood during harvest as the presence of impurities in honey hastens the fermentation process.

As honey is hygroscopic, readily absorbing moisture from the air, honey combs should be placed in clean, dry and hermetically closed containers. Squeezing honey combs should be avoided as it increases the risk of contamination by pollen or other impurities. Honey combs should preferably be cut in small cubes and left to drain over a piece of cheesecloth or a fine-mesh stainless steel strainer. Splitting the honey combs lengthwise and uncapping both sides eases the drainage of the honey (Figure 36).

Figure 36: Splitting the honey combs lengthwise and uncapping both sides eases the drainage of the honey. Photo: Eric Guerin.

As it is usually organic (forest honey) and harvested in a bee-friendly way, rafter honey can be sold at high price (Petersen 2012).

Depending on floral resources, a first harvest can take place 20 to 30 days after the bees settled on the rafter. And as the bees frequently rebuild the honey section, a second harvest can be carried out one month later (Ching, et al., 1995). It is usually possible to harvest two or three times from a colony before it absconds (Figure 37).

Figure 37: A rafter after two harvests have been taken (Vietnam). Source: Chinh et al., 1995.

At the end of the season the whole comb should be removed from the rafter to allow further colonization. Bees will not use an old comb thus bee wax can be harvested at this stage.

Rafter beekeepers commonly process wax by melting it and filtering it through a cheesecloth, a fine-mesh stainless steel strainer or coconut leaves (Figure 38). Melted wax is then slowly cooled.

Figure 38: An original wax filtration method using coconut leaves as a filter. Photo: Eric Guerin.

7. ACKNOWLEDGEMENTS

The author wishes to thank the rafter beekeepers from Cambodia, Thailand and Vietnam, Pieng Chhoin, Dani Jump, Dr. Phung Huu Chinh, Dr. Pham Hong Thai, Sanit Rattagan and Stephen Petersen for sharing their experience on rafter beekeeping. I would also like to thank Dr. Phung Huu Chinh and Jean-Pierre Chretien for the proof reading of this manual.

8. REFERENCES

Bees unlimited (https://www.beesunlimited.com).

Bendem-Ahlee, S., Kittitornkool,J., Thungwa, S., & Parinyasutinun, U. (2014). Bang Kad: A reflection of local wisdom to find wild honey and ecological use of resources in Melaleuca forest in the Songkhla lake basin. Silpakorn University Journal of Social Sciences, Humanities, and Arts, 14(3), 77–99.

Bendem-Ahlee, S., Kittitornkool, J., Thungwa, S., & Parinyasutinun, U. (2015). Honey hunter's way of life at Thung Bang Nok Ohk forest in the lower Songkhla lake basin amidst climate change in southern Thailand. Silpakorn University Journal of Social Sciences, Humanities, and Arts, 15(2), 105–126.

Chinh, P.H., Minh, N.H., Thai, P.H., and Tan, N.Q. (1995). Raftering - A traditional technique for honey and wax production from *Apis dorsata* in Vietnam. Article In Bees for Development Journal.

Chuttong, B., Somana, W. and Burgett, M. (2019). Giant Honey Bee (*Apis dorsata F.*) Rafter Beekeeping in Southern Thailand. Bee World, DOI: 10.1080/0005772X.2019.1596546.

Crane, E. (1990). Bees and Beekeeping: Science, Practice and World Resources. London: Heinemann Newnes.

Crane, E., Van Luyen, V., Mulder, V., and Ta, T. C. (1993). Traditional management system for *Apis dorsata* in submerged forests in southern Vietnam and central Kalimantan. Bee World, 74, 27–40. doi:10.1080/00057 72X.1993.11099151.

Crane, E. (1999). The World History of Beekeeping and Honey Hunting. pp 682.

Hadisoesilo, S. (2001). Tingku, a traditional management technique for *Apis dorsata* binghami in Central Sulawesi, in IBRA (eds.) proceedings of the 7th International conference on tropical bees management and diversity and the 5th Asian Apiculture Association conference, 19-25 March, 2000, Chiang Mai Thailand, International Bee Research Association, Cardiff, U.K. pp 309-312.

Ibrahim, I.F., Siva, Balasundram, S.K., Abdullah, N.P., Sood, A.M., Alias, M.S., Mardan, M., & Saberioon, M.M. (2012). The spatial distribution of Apis dorsta host plants using an integrated geographical information system- remote sensing approach. American Journal of Agricultural and Biological Sciences, 7(4), 396-406. doi:10.3844/ajabssp.2012.

Jack, C.J., Lucky, A., and Ellis, J.D. (2015). Entomology and Nematology Department, University of Florida. Photographs: Vereecken via Flickr; Sémhur via Wikimedia Commons; Nikolaus Koeniger, Martin–Luther- Universität, Institut für Biologie, Bereich Zoologie

Jump, D., & Waring, C. (2004). Rafter beekeeping in Cambodia. Bee Craft, 86(1), 4–5.

Koeniger, N., Koeniger, G., and Tingek, S. (2010). Honey Bees of Borneo: Exploring the Centre of *Apis* Diversity. Natural History Publications (Borneo), Kota Kinabalu, Sabah, Malaysia. Mahendra D.B. 1997. *Improved methods* of honey *harvest from rock bee colonies*, Indian Bee Jour. 1997. 59(2), 95-98.

Mahendre, D.B. (1997). Improved methods of honey harvest from rock bee colonies, Indian Bee Jour. 1997. 59(2),95-98.

Mulder, V., Heri, V., and Wickham, T. (2000). Traditional honey and wax collection with *Apis dorsata* in the Upper Kapuas Lake Region, West Kalimantan: a prelude to co-management. Published in Borneo Research Bulletin 31, 246-61

Nagir, M.T., Atmowidi, T., and Kahono, S. (2016). The distribution and nest-site preference of *Apis dorsata* binghami at Maros Forest, South Sulawesi, Indonesia. Journal of Insect Biodiversity Journal of Insect Biodiversity. 4(23), 1-14.

Nurtjahya, K. (2012). The Architecture of Sunggau in Banka Island, Indonesia – An Artificial wild Honey bee Nest, paper presented at 11th Asian Apicultural Association conference, Terengganu, West Malaysia 28th Sept- 2nd Oct. 2012.

Petersen, S. (2005). *Adventures in Beekeeping – Rafter Beekeeping in Cambodia.* American Bee Journal, Vol. 145, No. # 3, March 2005, pp. 222-224.

Petersen, S. (2010). *Rafter Beekeeping In Borneo.* American Bee Journal, Vol. 150, No. 5, May 2010, pp. 479- 483.

Petersen, S. (2012). Requirements for the Rafter Method of Sustainable Management with *Apis dorsata*.

Petersen, S. and Reddy, M.S. (2016). Requirements for Sustainable Management of *Apis dorsata* Fab. with Rafter Method. In book: Arthropod Diversity and Conservation in the Tropics and Sub-tropics, pp.383-396. DOI: 10.1007/978-981-10-1518-2_24

Purwanto, D. B., Hadisoesilo, S., Kasno, Koeniger, N., Lunderstedt, J. (2001). *Sunggau system a sustainable method of honey production from Indonesia with the giant Asian Honey bee Apis dorsata*, in IBRA (eds.) proceedings of the 7th International conference on tropical bees management and diversity and the 5th Asian Apiculture Association conference, 19-25 March, 2000, Chiang Mai Thailand, International Bee Research Association, Cardiff, U.K.; pp. 201-206.

Reddy, M.S. & Reddy, C.C. (1989) Height - dependent nest site selection in *Apis dorsata* Fabr. Indian Bee Journal 51, 105-106.

Ruttner, F. (1988) Biogeography and Taxonomy of Honey bees. Berlin: Springer-Verlag.

Saberioon, M.M., Mardan, M., Nordin, L., Alias, M.S., and Gholizadeh, A. (2010). Predict Location(s) of *Apis dorsata* Nesting Sites Using Remote Sensing and Geographic Information System in Melaleuca Forest. Journal of Apicultural Research.

Sihag, R.C. (2017). Nesting behavior and nest site preferences of the giant honey bee (*Apis dorsata* F.) in the semi-arid environment of north west India, Journal of Apicultural Research, DOI: 10.1080/00218839.2017.1338443

Tan, N.Q. (2004). Studies of the Asian giant honey bee, *Apis dorsata Fabricius* (Apidae) in the submerged melaleuca forest of Vietnam: biology, behavior, ecology and apiculture. Wolfson College Thesis submitted for the Degree of Doctor of Philosophy in the Department of Zoology, University of Oxford

Tan, N.Q., Chinh, P.H., Thai, P.H., & Mulder, V. (1997). Rafter beekeeping with *Apis dorsata*: some factors affecting the occupation of rafters by bees. Journal of Apicultural Research, 36(1), 49–54. doi:10.1080/00218839.1997.11100930

Thakar, C. (1973). A Preliminary Note on Hiving *Apis dorsata* Colonies. Bee World. 54. 24-27. DOI: 10.1080/0005772X.1973.11097444.

Waring, C., & Jump, D. (2004). Rafter beekeeping in Cambodia with *Apis dorsata*. Bee World, 85(1), 14–18. doi:10.1080/0005772X.2004.11099607

Woyke, J. (2017). Curtains, garlands and face of arriving *Apis dorsata* swarms. Warsaw University Of Life Sciences.

www.ingramcontent.com/pod-product-compliance
Lightning Source LLC
Chambersburg PA
CBHW060900090426
42738CB00022B/3482